YOUR KNOWLEDGE HAS VALUE

- We will publish your bachelor's and master's thesis, essays and papers

- Your own eBook and book - sold worldwide in all relevant shops

- Earn money with each sale

Upload your text at www.GRIN.com and publish for free

Bibliographic information published by the German National Library:

The German National Library lists this publication in the National Bibliography; detailed bibliographic data are available on the Internet at http://dnb.dnb.de .

Imprint:

Copyright © 2018 GRIN Verlag
Print and binding: Books on Demand GmbH, Norderstedt Germany
ISBN: 9783668817517

Moritz Stüber

Conventional and Organic Coffee Plantations and their Effects on Arthropods and Avifauna. A Biodiversity Check in Costa Rica

GRIN Verlag

DIVISION OF ORGANIC FARMING
WG TRANSDISCIPLINARY SYSTEMS RESEARCH

Individual homework

Conventional and organic coffee plantations and their effects on arthropods and avifauna - A biodiversity check in Costa Rica

Name of the course: EUR-Organic Start-Up Module

Name: Moritz Stueber

Current University: Hohenheim University

Due date to upload the final version: 28th of February 2018

The report counts **25'231** characters, excluding: (6) References, (7) Annex, direct quotation, figures and illustrations nor titles.

Table of content

List of tables

List of figures

List of abbreviations

m a.s.l.	metres above sea level
Ca	Calcium
CELAC	Community of Latin American and Caribbean States
CITES	Convention on International Trade in Endangered Species of Wild Fauna and Flora
CLS	coffee leaf scorch
CR	Costa Rica
e.g.	exempli gratia (for example)
f.e.	for example
ff	and the following
ha	hectare
HIVOS	Humanist Institute for Development Cooperation
ICA	International Coffee Agreement
IFOAM	International Federation of Organic Agriculture Movements
IUCN	International Union for Conservation of Nature
IUCN Red List	IUCN Red List of Threatened Species
km	kilometres
Mg	Magnesium
Mio.	Million
OA	Organic Agriculture
OECD	Organisation for Economic Co-operation and Development

Glossary

BF	Smithsonian Bird Friendly (Certification)
Cam(Bio)2	Agricultural initiative of the CEDECO ("Agricultores del clima")
CEDECO	Educative Corporation of Development in Costa Rica ("Corporación Educativa para el Desarrollo Costarricense")
Fairtrade	Certification body on international trade
FONECAFE	National Fund for Coffee Stabilisation ("Fondo Nacional de Estabilización Cafetalera")
ICAFE	Coffee Institute of Costa Rica ("Instituto del Café de Costa Rica")
MAG	Ministry of Agriculture and Livestock ("Ministerio de Agricultura y Ganadería de Costa Rica")
MINAE	Ministry of Environment, Energy and Telecommunications ("Ministerio de Ambiente y Energía de Costa Rica")
Neotropics	Neotropical realm, including South and North American regions of Central America

Specification of geographical indications in Costa Rica can be found in the annex (Table 1)

1 Introduction

1.1 Introduction

When it comes to ecological measures and sustainability, Costa Rica (CR) is one of the most developed countries in Latin America. During the past two years CR was able to be nearly 100% eco-power self-sufficient (MINAE 2015). Moreover it is planning to be the first country worldwide to prohibit disposable plastic (Soto 2016) and one of the organic pioneers in the region. Organic food export and people's ecologic awareness have been rising (Barquero 2013). As the biodiversity in CR is one of the highest worldwide, the country's nature protection strategy is a topic to focus on and worth to be studied (Obando-Acuña 2007). Latin American studies on organic management as a tool for biodiversity conservation are rare (Foguelman 2005, p.87–93). Organic Agriculture (OA), as written in the IFOAM 1972, always committed to the conservation of biodiversity. There were only few attempts to include biodiversity conservation into the existing organic standards and guidelines. First in 2005, biodiversity conservation, as part of the draft "Biodiversity and Landscape Standards" was mentioned and integrated into the IFOAM Basic Standards (Stolton 2005, p.10ff).

In CR, one of the principal promoters of OA is the CEDECO (Educative Corporation of Development in Costa Rica). Their research often focused on arthropods since many species are known to be biological indicators of an ecosystem's health status. Exemplary insects of their studies are butterflies or dung beetles (CEDECO 2010; Sánchez et al. 2014). Many birds are insect-/ omnivorous, thus directly linked to arthropod abundance. Organic coffee plantations in CR are tried to be integrated into nature (agroforestry), which results in a different habitat than the conventional ones. Instead of shrub-sized coffee plants, there is a sudden interaction with the canopy and a more diverse shrub-layer (Florian Rivero 2005, p.10-13, 80ff). In the following report, the focus will be laid on species that experienced a success story but also those who might not. It will be demonstrated by changes in vegetation, diversity changes above and beneath ground and the way the farmer can profit from each species.

1.2 Research questions

1) Which species especially thrive on organic coffee plantations? What adverse effects on species can be seen in organic coffee cultivation/ coffee agroforestry?

2) What are the main factors of an organic coffee plantation to increase its biodiversity?

3) From a farmer's point of view, what benefits does a biodiverse organic coffee plantation provide?

1.3 Objectives

Intensive agriculture and the use of chemicals have been known as one of the main causes of biodiversity losses. OA contributes to a solution of concurrent agriculture and nature conservation. Case studies under local conditions are to be investigated, to prove the potential of organically managed land. These investigations often lack an overall structure and prioritise different aspects, it is therefore challenging to compare and later merge them together into one clear statement. This report aims to summarise the latest studies for a specific cultivation (coffee) in a certain area (CR). The health status of the ecosystem is analysed by means of diversity and abundance of arthropods and bird species as their direct consumers. The different approaches of organic and conventional agriculture are tried to be drawn together with the kind of species that may have experienced positive or negative effects after an organic conversion. The reasons behind these effects are to be analysed and eventually formed into a defined recommendation of biodiversity favouring actions. Furthermore, the nature conservation and the farmer's point of view and benefits will be included within the report. This report will eventually lead to a clear answer of the three research questions. It will therefore be able to clearly point out the potential of biodiversity conservation in Central American coffee cultivation, potential mutual benefits and defined recommendations on biodiversity conservation.

2 Methods

The report is a literature review, mainly based on sources found at research databases, such as "ScienceDirect" or "ResearchGate". The literature research was performed in English and Spanish. The following terms were used throughout the research: *"biodiversity; organic agriculture; agroforestry; Costa Rica; coffee"* and further detailed scientific terms, such as *"arthropods"* or *"Coleoptera"*. Literature was also found at local libraries. The author furthermore addressed local contacts and organisations to gain more knowledge about consisting structures and case studies.

3 Results

3.1 Overview of organic coffee production in Costa Rica

Costa Rica, as one of the most prosperous Central American CELAC-countries stands for coffee of high-quality. Coffee is mainly cultivated in the Central Valley and the Central/ Southern Pacific region (Figure 1; orange frame), 92% of production can be found within this area. 41% of the production is located only in the province of San José (OECD 2017). It is the high elevation of the Central and Talamanca mountain ranges, surrounding the Central Valley, that favours the coffee production in this region. From 1'200-1'500 m a.s.l. in the Central Valley (Figure 1, red frame) cities San José (capital) and Cartago, the elevation rises to 3'500 m a.s.l. within 10-20 km. Over the past thirty years, coffee production shrank from a market share of 14 to only 6% of the CR agricultural sector. It is still the third biggest arable crop, right after pineapple (24%) and banana (18%). Pineapple production increased the most during that period, at the expense of coffee and banana production (OECD 2017).

Figure 1: Map of Costa Rica, (orange frame: main coffee production in CR; red frame: Central Valley; the surrounding mountain ranges indicated), further explanation in the annex (Table 1). Source: https://commons.wikimedia.org/wiki/Atlas_of_Costa_Rica

The coffee sector in CR regulated by the semi-autonomous CR Coffee Institute (ICAFE), protecting quality and reputation. Regulative measures are e.g. retaining 2% of the coffee with the lowest quality from the export market (Varangis et al. 2003) or protecting producers by limiting expenses of coffee milling and banning farmgate retailing activities (Icafe 1961, p.2ff). Responding to the crisis in 1989 after the collapse of the ICA (International Coffee Agreement), CR implemented the "Fund for Coffee Stabilization" (FONECAFE) to strengthen financial security of the coffee production sector (Varangis et al. 2003). What on the one hand leads to income security and compensation for world price drops, leads on the other hand to an expensive production and a highly standardised product.

The requirements for certified coffee (e.g. organic coffee) coincide with the agriculture in CR, thanks to the environmental and social laws and the coffee sector regulations. Though one of the organic pioneers in the region, organic production in CR accounts for 1.6% (8'000 ha in 2016) of the total agricultural production. The organic market share is lower than in other adjacent countries and below global average. Now organic products are on the rise again, after several years of fluctuating prices, high production (labour) costs and difficulties of receiving price premiums (MAG 2016; OECD 2017). Unfortunately, organic coffee in CR poses a far greater challenge, than the framework conditions might suggest (Snider et al. 2017). The CR coffee production is intensive, especially concerning the inputs. A conversion to organic coffee production results in a 50% lower yield than in conventional production. Adjacent countries like Nicaragua or Guatemala have a comparative advantage. Compared to these countries, the CR coffee production is at a relatively low elevation. Coffee production in these countries is less intensive and thus not confronted with a yield loss, too great to bear after the organic conversion. For big corporations like Starbucks or Nespresso, offering their own certified coffee, these countries are more favourable. The CR coffee sector needs certification bodies like Fairtrade or Organic, to add value and increase its demand (Snider et al. 2017). Organic coffee accounts for 9% of the country's coffee production, more than in conventional (MAG 2016). 31% of the organic sector are meant for the domestic market, 69% are meant for export (MAG 2013). Still there is no national strategy to simplify the small-scale farmer's access to the global market.

3.2 Biodiversity in Costa Rica

The biodiversity in CR belongs to the highest worldwide. Within an area smaller than Bavaria, it accounts over 90'000 (2 Mio. presumed) species in 2005. In absolute terms, it belongs to the 20 megadiverse countries. CR has the highest density of species in Central America and maybe even worldwide. The country owes its species richness to its geographical position in the Neotropics (tropical zone of the American continent) and its geological history as part of the bridge between the two continental land masses. The landscape is mountainous, which results in thousands of different microclimates. CR has access to two different costs (Pacific and Atlantic/Caribbean). Around 25% of the country's national territory are protected areas. One of the hotspots of protection are the mountain ranges, surrounding the Central Valley, which is simultaneously one of the hotspots of coffee production (Obando-Acuña 2007).

Next to the international organization IUCN (Red List) and CITES, the MINAE ("Ministry of Environment, Energy and Telecommunications") counts as one of the most important conservation bodies. The ministry's official decree 32633-MINAE (MINAE 2005, p.11–14) contains a list of endangered species. The list consists of around 1'600 of the 90'000 proved species, considering mainly plants and vertebrates. Most invertebrates are excluded. Therefore, only ~2% of all species known are registered as endangered or threatened by extinction, with increasing tendency. The ratio of endangered species increases, by looking more specifically. 10% of all vertebrates and 12% of all plants are considered endangered/ threatened by extinction. Of birds, 11.5% are endangered (Obando-Acuña 2007).

3.3 Biodiversity check

Agriculture, located in a biodiversity hotspot like the CR Central Valley automatically competes for space against tropical habitats and protected areas. Agricultural land is favourable for tropical crops, like coffee, thanks to the climate conditions and fertile land. Available agricultural land remains scarce (OECD 2017). The two systems might be linked together by agroforestry. Agroforestry is broadly defined as a land-use system of woody perennials (trees, shrubs) used together with crops or livestock in the same area. The two different components are interacting ecologically and economically. It can be also described as a *"dynamic, ecologically based, natural resource management system that, through the integration of trees on farms and in the agricultural landscape, diversifies and sustains production for increased social, economic and environmental benefits for land users at all levels"* (FAO 2015).

One of the archetypes of organic and biodynamic coffee agroforestry, "Finca Irlanda" in Chiapas (Mexico), is growing more than 40 varieties of leguminous trees, providing shade and nitrogen. Several species that are threatened by extinction, returned to the plantation, e.g. the pum (*Puma concolor*). By settling ocelots or grey foxes, the organic agroforestry scheme also led to the control of pests, e.g. rodents (Stolton 2002, p.9-14). The farm, that was later criticised for its inhuman working conditions (Bringmann 1997), successfully restored biodiversity at its 300 ha plantation (Perfecto et al. 2003).

3.3.1 Arthropods

Many insects are seen as health indicators. E.g., they have an active role in pollination, decomposition and soil maintenance. They can be very beneficial to a farmer, for example in pollinating crop plants or input nitrogen. Many case studies and biodiversity checks are therefore focussing on arthropods. Within a research on OA and its impact on climate change, the CEDECO together with its agricultural initiative "Cam(Bio)2", performed a biodiversity check of organic and conventional coffee plantations in Caraigres, San José (close to the Central Valley). They focused on dung beetles and butterflies, because of their status as ecological health indicators (CEDECO 2010). Especially the abundance of dung beetles (genus *Onthophagus* and *Eurysternus*) and butterflies (genus *Cissia* and *Eurema*) tremendously increased at organically managed plantations. Differences in the diversity of dung beetles were insignificant. Conventional coffee plantations showed a higher butterfly diversity. The authors found 33 butterfly species, unique to the conventional system and only 19 butterfly species for the organic system. Non-frugivorous butterfly species (e.g. nectivorous) are more common in open areas, butterfly diversity might therefore be increased at conventional coffee plantations. The authors argued, that a higher abundance of butterflies came from the presence of herbs and trees in organic plantations. The absence of chemical herbicides and intensive management is favouring diverse plant growth and dung beetle abundance. Conventional coffee plantations are often managed in open areas, whereas organic coffee plantations, e.g. as agroforestry, are kept more naturally balanced within the tropical forest ecosystem (Sánchez et al. 2014). Flowering plants like epiphytes on top of the trees are also attracting pollinators. The bee diversity of CR coffee farms is higher if flowering plants are integrated into the borders. Bee pollination has also been shown to increase the yields in coffee production. A conversion to organic coffee plantations can also make ant abundance increase around 20% (Rice and Bedoya 2010).

George (2006, p.49-52) were analysing the soil quality of organic and conventional coffee plantations. Inter alia, the authors were focussing on collembolan, nematode and earthworm populations. Nematode and earthworm abundance were positively correlated with organic management. Collembolan and nematode populations had an increased diversity on organically managed coffee farms. Conventional and un-shaded coffee plantations showed a lower abundance and diversity. The attributes that favoured diversity and abundance in organic coffee plantations were higher soil pH-value, Mg & Ca concentration and a higher nitrogen availability, thanks to a lower carbon/nitrogen ratio.

3.3.2 Birds

Birds are often insect-/ omnivorous and therefore depend on insect abundance. As many insects are health indicators, bird abundance is a further indication of a healthy ecosystem. The "Smithsonian Bird Friendly" (BF) certification by the "Migratory Bird Center", as part of the "Smithsonian's National Zoo and Conservation Biology Institute", is the first scientifically approved shade-grown coffee certification body worldwide. All labelled coffee plantations also have to be certified Organic. Coffee grown under a canopy is offering shade and by that a semi-artificial intermediate between agriculture and nature conservation. In their 2010 report Rice and Bedoya (Smithsonian Migratory Bird Center) claim that a higher tree abundance as shade cover in a coffee plantation generally leads to a higher biodiversity. There are still missing representative case studies, comparing BF and non-BF farms. Generally, taller and structurally more diverse canopies lead to a higher biodiversity. The few head-to-head comparisons of BF (organic) and non-BF plantations showed around 20% more bird diversity at BF plantations. The presence and abundance of ants counts as an indicator of habitat health and natural pest control. Coffee plantations with integrated fruits might have a mutual value to farmer and biodiversity. CR coffee farmers often plant fruit trees in their plantations. The trees can be used economically or work stimulating, by providing a little snack during the harvest season. In either way, birds can benefit by gaining a valuable resource. These fruit trees can attract birds and double the bird species diversity. Furthermore, trees in the tropics often harbour epiphytes, like bromeliads and orchids. The plants are often used as nesting material. Insects, attracted by the flowers, are a valuable food source for birds (Rice and Bedoya 2010).

Figure 2: *E. Poeppigiana* (left) & *C. alliodora* (right) used in agroforestry coffee systems during the study of Florian (2005).

Source: https://fr.m.wikipedia.org/wiki/Fichier:Erythrina_poeppigiana_(18258348879).jpg

https://commons.wikimedia.org/wiki/File:Cordia_alliodora.jpg

Florian Rivero (2005, p.10-13, 80ff) came to similar results in the study of bird diversity in coffee agroforestry systems in Turrialba in the Central mountain range. The author was comparing the diversity of tropical birds in conventional coffee plantations and different types of agroforestry system. A higher structural complexity of the plantation is favouring bird abundance and diversity. Coffee agroforestry systems, using coffee crops and a legume plant (*Erythrina poeppigiana*) had a lower species richness than systems, that were using two extra plants (*E. poeppigiana* and *Cordia alliodora*) (Illustration 2). The observed birds throughout the study were mainly insectivores and omnivores. Epiphyte abundance for cover, the canopy height and the number of *C. alliodora* plants had a positive correlation with bird diversity. Negatively correlated with the species richness were surrounding dense forests but were on the other hand favouring the abundance of forest specialising birds. The author promoted the diversification of coffee farms and protection of the surrounding forest, in order to conserve biodiversity and to maintain a structural and functional landscape connectivity.

4 Discussion

The organic sector in CR has a huge potential, thanks to rising welfare standards and a rising demand for certified products (Barquero 2013). Though having a mainly export oriented agriculture, the conventional but especially the organic market lack structure and production chains (MAG 2013; MAG 2016). However, an increase in organic production could have a major beneficial impact on biodiversity conservation. In a small country like CR, protected areas and agriculture are naturally competing for space. Coffee in CR is the perfect example because it is mainly cultivated in the biodiversity hotspot of the humid elevations around the Central Valley. Therefore, the focus moves to nature integrating systems like agroforestry. Managed by certification bodies like Organic, plus the absence of agrochemicals, agroforestry might be the ideal system of modern tropical agriculture. The few studies that were undertaken on that specific subject showed astounding increases in biodiversity. The diversity and abundance of insects and other arthropods, health indicators of natural systems and birds (feeding on arthropods) were increasing parallel to a rising structural complexity of organic agroforestry systems. By changing from two plants (coffee + legume tree) to a threefold agroforestry system (coffee + legume tree + shrub) the agroforestry system in the case study of Florian Rivero (2005, p.10-13, 80ff) tremendously increased bird diversity and abundance. Furthermore, tree height and absence of chemical pesticides are diversity favouring mechanisms. Coffee plants are mainly kept in shrub size and rather count as an open area than a forest. For open area adapted species, like nectivorous butterflies, agroforestry means a reduction in abundance (Sánchez et al. 2014). Also forest specialist bird species (Florian Rivero 2005, p.10-13, 80ff) are increasing, if the surrounding forest is too dominating. The overall bird diversity will then decrease, the forest specialist diversity increase. Organic coffee plantations can also be beneficial to the farm management. A higher abundance of predators (ocelots, ants) as natural pest control (rodents, grasshoppers) and detritivores (earthworms, nematodes) will benefit the habitat health and simultaneously the crop and soil conditions. Wild bees can be often located at coffee plantations. A higher wild bee abundance in agroforestry systems can have a major influence on the pollination service and yield increase (Perfecto et al. 2003; Florian Rivero 2005, p.10-13, 80ff; Rice and Bedoya 2010). Fruit trees can also be part of a coffee plantation. A measure, that farmers often implement culturally. Integrated fruit trees have a value to birds and insects, but also to the farmers themselves. Whether the fruits are used as an economic by-product or just for domestic consumption, the cultivation is still benefiting farmer and nature. Besides their function as ecosystem health indicators, some insects are seen as pests. Sharpshooters of the family Cicadellidae f.e. are potential vectors of *Xylella fastidiosa*, the bacteria of the coffee disease "coffee leaf scorch" (CLS). In the Talamanca mountain range, organic (agroforestry) coffee plantations include shade and structural management. Most sharpshooters are more aligned to pastures and open areas than forests, what makes conventional plantations vulnerable. Agroforestry systems, containing *E. Poeppigiana* & *C. alliodora* (Illustration 2) have the locational advantage and therefore a lower abundance of sharpshooters. In this case, organic agroforestry can even lead to a higher disease resistance (Ramos 2008, p.22-45).

5 Conclusion

Despite its potential of being a leading force and role model of Central American organic production, CR's organic market is still under development. CR was a pioneer of the Central American organic market. Unfortunately, the market share of organic products is currently below global average and below comparable, adjacent countries like Nicaragua or Guatemala. One of the major problems in a small but nature diverse country like CR is the alignment of agriculture and biodiversity conservation. Especially in biodiversity hotspots, a sustainable agriculture is necessary. OA and agroforestry might be a good solution to combine the two opposing systems. In the case of coffee plantations, a conversion to organic agroforestry led to an increase of biodiversity, especially of ecosystem health indicating species. The overall arthropod abundance seemed to increase, as well as the diversity of birds. Only in rare cases of a to intensive forest integration, open area species like some birds and nectivorous butterflies were declining. The benefiting arthropods are also beneficial to the farmer, thanks to their predatory habits (ants), pollination (bees, butterflies) or decomposition (dung beetles, earthworms) (Florian Rivero 2005, p.10-13, 80ff; Rice and Bedoya 2010; Sánchez et al. 2014). A higher abundance of bees eventually even leads to a higher yield in coffee production. Even though most of the results found during the report were positive, the lack of an appropriate number of encompassing studies makes the few positive investigations seem indefinite. The reason for missing representative case studies are difficulties of finding comparable plantations. The extraordinary number of microclimates in the mountain ranges, thanks to climate conditions, different elevation types and soil quality, makes it hard to compare different farms.

6 References

Barquero, M. (10.07.2013): Nuevos negocios se especializan en vender productos orgánicos en Costa Rica, http://www.nacion.com/economia/Nuevos-especializan-organicos-Costa-Rica_0_1352864762.html. Last access 29.10.2017.

Bringmann, R. (01.01.1997): Finca Irlanda – Erfolg der Boykott-Kampagne (Rücknahme des "FairTrade"-Siegels von Lebensbaum), http://lateinamerika-nachrichten.de/?aaartikel=finca-irlanda-erfolg-der-boykott-kampagne. Last access 23.02.2018.

CEDECO (2010): Agricultura orgánica y cambio climático, Revista Aportes, Edición N° 137, 11, 36-37.

FAO (23.10.2015): Agroforestry (Definition), http://www.fao.org/forestry/agroforestry /80338/en/. Last access 23.02.2018.

Florian Rivero, E. Marcela (2005): Tropical bird assemblages in coffee agroforestry systems: exploring the relationships between landscape context, structural complexity and bird communities in the Turrialba - Jiménez Biological Corridor, Costa Rica, University of Idaho, Idaho, 10-13, 80ff.

Foguelman, D. (2005): Biodiversity: A Target in Organic Latin America, 87–93.

George, A. (2006): Estudio comparativo de indicadores de calidad de suelo en fincas de café orgánico y convencional en Turrialba, Costa Rica, Turrialba, Costa Rica, 49-52.

Icafe (1961): Régimen Relaciones de Productores, Beneficiadores y Exportadores Café, 2ff.

MAG (2013): Estudio sobre el entorno nacional de la agricultura orgánica en Costa Rica (Programa Nacional de Agricultura Orgánica), 40.

MAG (2016): Servicio Fitosanitario del Estado (Acreditación y Registro de Agricultura Orgánica), 2-3.

MINAE (2005): Reglamento a la Ley de Conservación de la Vida Silvestre.

MINAE (2015): Plan Nacional de Energía 2015-2030, Ministerio de Ambiente y Energia, San José, Costa Rica, 1-2.

Obando-Acuña, V. (2007): Biodiversidad de Costa Rica en cifras, Instituto Nacional de Biodiversidad (INBio), Costa Rica, 8-17.

OECD (2017): Agricultural Policies in Costa Rica, OECD Publishing.

Perfecto, I., A. Mas, T. Dietsch, J. Vandermeer (2003): Conservation of biodiversity in coffee agroecosystems: a tri-taxa comparison in southern Mexico, Kluwer Academic Publishers, 12, 1239–1252.

Ramos, M. (2008): The effects of local and landscape context on leafhopper (Hemiptera: Cicadellinae) communities in coffee agroforestry systems of Costa Rica, University of Idaho, Idaho, 22-45.

Rice, R., M. Bedoya (2010): The Ecological Benefits of Shade-Grown Coffee, https://nationalzoo.si.edu/scbi/migratorybirds/coffee/bird_friendly/ecological-benefits-of-shade-grown-coffee.cfm. Last access 23.02.2018

Sánchez, N. V., L. E. Vargas-Castro, A. Sánchez, M. Amador (2014): Riqueza y abundancia de mariposas diurnas, escarabajos coprófagos y plantas en cultivos orgánicos y convencionales de tres regiones de Costa Rica, UNED Research Journal, vol. 5, 249–257.

Snider, A., I. Gutiérrez, N. Sibelet, G. Faure (2017): Small farmer cooperatives and voluntary coffee certifications, Food Policy, 235–236.

Soto, M. (16.11.2016): Costa Rica se encamina a eliminar los plásticos de un solo uso, www.nacion.com/vivir/ambiente/Costa-Rica-encamina-eliminar-plasticos_0_1597840214.html. Last access 29.10.2017.

Stolton, Sue (2002): Organic agriculture and biodiversity, IFOAM, Imsbach.

Stolton, Sue (2005): Organic agriculture for biodiversity (Current contributions and future possibilities; proccedings of the Third International IFOAM Conference on Biodiversity and Organic Agriculture, Nairobi, Kenya 2004), IFOAM, Imsbach, 10ff.

Varangis, P., P. Siegel, D. Giovannucci, B. Lewin (2003): Dealing with the Coffee Crisis in Central America, The World Bank, Washington, DC.

7 Annex

Table 1: Specification of regional geographical terms

Term	Specification
Caraigres	Region in the province San José, part of the Talamanca mountain range
Cartago	City in the Central Valley
Central Valley	Geographical region with the largest urban agglomeration in Costa Rica, the elevations onto the surrounding mountain ranges count as the biodiversity hotspots in Costa Rica
Central mountain range	Mountain range in the northwest of the Central Valley
San José	Capital, city in the Central Valley
Talamanca mountain range	Mountain range in the southeast of the Central Valley
Turrialba	City and region, famous for its volcanic activity (volcano "Turrialba")

ABSTRACT

The subject of this report was the investigation of differences in conventional and organic management of coffee plantations and potential benefits for biodiversity. It is mainly based on literature researches and case studies of the past 20 years. Biodiversity hotspots (and conservation) and the coffee industry are mainly located in the same region in the elevations of the Central Valley. Costa Rica is therefore the optimal example, since a conscious and sustainable agriculture is the only way to maintain biodiversity. The three main subjects of this report were the overall diversity change after a conversion, the reasons behind the diversity changes and potential benefits of a conversion for a coffee farmer. To look at the health status of an ecosystem, the report mainly focused on arthropods and their direct consumers (birds). In all case studies, the biodiversity increased after the conversion to Organic. The overall abundance of birds and insects on the plantations were increased. Especially dung beetles, bees and bird species experienced a success story. Some butterfly species, that were specialised on open areas did not benefit from the conversion. However, an overall biodiversity increase could be proved. The crucial measure to increase biodiversity is structural complexity of the agroforestry system. Measures that favoured biodiversity in coffee agroforestry systems are the height of trees, harboured epiphytes (e.g. orchids) on trees and especially the number of different plant species. After adding a third plant (*C. alliodora*) to the agroforestry system, biodiversity increased tremendously (Florian Rivero 2005, p.10-13, 80ff). The farmer can gain crucial benefits from an increased biodiversity on the plantation. Wild bees are proved to increase the yield of coffee plants (Rice and Bedoya 2010), but also an increased abundance of overall pollinators is benefiting the cultivation. Decomposition and nitrogen ratio in the soil are increasing, thanks to a higher number of detritivores (George 2006, p.49-52). Organic agroforestry is also attracting predators (e.g. ants) that commit to natural pest control. The research questions have been answered in favour of OA. Unfortunately, the number of varying microclimates in the Costa Rican mountain region are making it difficult to find two comparable organic and conventional farms. There is still missing a valuable number of case studies to clearly indicate organic to be superior to conventional coffee production, concerning nature conservation.

References

Barquero, M. (10.07.2013): Nuevos negocios se especializan en vender productos orgánicos en Costa Rica, http://www.nacion.com/economia/Nuevos-especializan-organicos-Costa-Rica_0_1352864762.html. Last access 29.10.2017.

Bringmann, R. (1997): Finca Irlanda – Erfolg der Boykott-Kampagne, http://lateinamerika-nachrichten.de/?aaartikel=finca-irlanda-erfolg-der-boykott-kampagne. Last access 23.02.2018.

CEDECO (2010): Agricultura orgánica y cambio climático, Revista Aportes, Edición N° 137, p.11, 36-37.

FAO (23.10.2015): Agroforestry, http://www.fao.org/forestry/agroforestry/80338/en/. Last access 23.02.2018.

Florian Rivero, E. Marcela (2005): Tropical bird assemblages in coffee agroforestry systems: exploring the relationships between landscape context, structural complexity and bird communities in the Turrialba - Jiménez Biological Corridor, Costa Rica, University of Idaho, Idaho.

Foguelman, D. (2005): Biodiversity: A Target in Organic Latin America.

George, A. (2006): Estudio comparativo de indicadores de calidad de suelo en fincas de café orgánico y convencional en Turrialba, Costa Rica, Turrialba, Costa Rica.

Icafe (1961): Régimen Relaciones de Productores, Beneficiadores y Exportadores Café.

MAG (2013): Estudio sobre el entorno nacional de la agricultura orgánica en Costa Rica, p.40.

MAG (2016): Servicio Fitosanitario del Estado, p.2-3.

MINAE (2005): Reglamento a la Ley de Conservación de la Vida Silvestre.

MINAE (2015): Plan Nacional de Energía 2015-2030, Ministerio de Ambiente y Energia, San José, Costa Rica, p.1-2.

Obando-Acuña, V. (2007): Biodiversidad de Costa Rica en cifras, Instituto Nacional de Biodiversidad (INBio), Costa Rica, p.8-17.

OECD (2017): Agricultural Policies in Costa Rica, OECD Publishing.

Perfecto, I., A. Mas, T. Dietsch, J. Vandermeer (2003): Conservation of biodiversity in coffee agroecosystems: a tri-taxa comparison in southern Mexico, Kluwer Academic Publishers, 12, p.1239–1252.

Ramos, M. (2008): The effects of local and landscape context on leafhopper (Hemiptera: Cicadellinae) communities in coffee agroforestry systems of Costa Rica, University of Idaho, Idaho.

Rice, R., M. Bedoya (2010): The Ecological Benefits of Shade-Grown Coffee.

Sánchez, N. V., L. E. Vargas-Castro, A. Sánchez, M. Amador (2014): Riqueza y abundancia de mariposas diurnas, escarabajos coprófagos y plantas en cultivos orgánicos y convencionales de tres regiones de Costa Rica, UNED Research Journal, vol. 5, p.249–257.

Snider, A., I. Gutiérrez, N. Sibelet, G. Faure (2017): Small farmer cooperatives and voluntary coffee certifications, Food Policy, p.235–236.

Soto, M. (16.11.2016): Costa Rica se encamina a eliminar los plásticos de un solo uso, www.nacion.com/vivir/ambiente/Costa-Rica-encamina-eliminar-plasticos_0_1597840214.html. Last access 29.10.2017.

Stolton, S. (2002): Organic agriculture and biodiversity, IFOAM, Imsbach.

Stolton, S. (2005): Organic agriculture for biodiversity (Current contributions and future possibilities ; proccedings of the Third International IFOAM Conference on Biodiversity and Organic Agriculture, Nairobi, Kenya 2004), IFOAM, Imsbach.

Varangis, P., P. Siegel, D. Giovannucci, B. Lewin (2003): Dealing with the Coffee Crisis in Central America, The World Bank, Washington, DC., p.17ff.

YOUR KNOWLEDGE HAS VALUE

- We will publish your bachelor's and
  master's thesis, essays and papers

- Your own eBook and book -
  sold worldwide in all relevant shops

- Earn money with each sale

Upload your text at www.GRIN.com
and publish for free